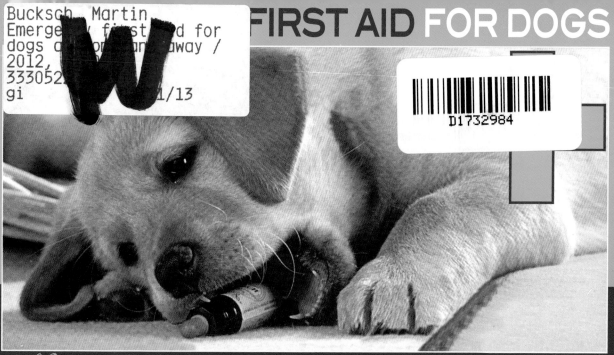

FIRST AID FOR DOGS

Hubble & Hattie

At home and away

The Hubble & Hattie imprint was launched in 2009 and is named in memory of two very special Westies owned by Veloce's proprietors. Since the first book, many more have been added to the list, all with the same underlying objective: to be of real benefit to the species they cover, at the same time promoting compassion, understanding and co-operation between all animals (including human ones!) Hubble & Hattie is the home of a range of books that cover all-things animal, produced to the same high quality of content and presentation as our motoring books, and offering the same great value for money.

More titles from Hubble & Hattie

A Dog's Dinner (Paton-Ayre)
Animal Grief: How animals mourn (Alderton)
Bramble: the dog who wanted to live forever (Heritage)
Cat Speak: recognising & understanding behaviour (Rauth-Widmann)
Clever dog! Life lessons from the world's most successful animal (O'Meara)
Complete Dog Massage Manual, The – Gentle Dog Care (Robertson)
Dieting with my dog: one busy life, two full figures ... and unconditional love (Frezon)
Dinner with Rover: delicious, nutritious meals for you and your dog to share (Paton-Ayre)
Dog Cookies: healthy, allergen-free treat recipes for your dog (Schöps)
Dog-friendly Gardening: creating a safe haven for you and your dog (Bush)
Dog Games – stimulating play to entertain your dog and you (Blenski)
Dog Speak: recognising & understanding behaviour (Blenski)
Dogs on Wheels: travelling with your canine companion (Mort)
Emergency First Aid for dogs: home and away (Bucksch)
Exercising your puppy: a gentle & natural approach – Gentle Dog Care (Robertson & Pope)
Fun and Games for Cats (Seidl)
Groomer's Bible, The The definitive guide to the science, practice and art of dog grooming for students and home groomers (Gould)
Know Your Dog – The guide to a beautiful relationship (Birmelin)
Life Skills for Puppies: laying the foundation for a loving, lasting relationship (Zulch & Mills)
Miaow! Cats really are nicer than people! (Moore)
My dog has arthritis – but lives life to the full! (Carrick)
My dog is blind – but lives life to the full! (Horsky)
My dog is deaf – but lives life to the full! (Willms)
My dog has hip dysplasia – but lives life to the full! (Haüsler)
My dog has a cruciate ligament injury – but lives life to the full! (Haüsler)
Older Dog, Living with an – Gentle Dog Care (Alderton & Hall)
Partners – Everyday working dogs being heroes every day (Walton)
Smellorama – nose games for dogs (Theby)
Swim to recovery: canine hydrotherapy healing – Gentle Dog Care (Wong)
The Truth about Wolves and Dogs: dispelling the myths of dog training (Shelbourne)
Waggy Tails & Wheelchairs (Epp)
Walking the dog: motorway walks for drivers & dogs (Rees
Walking the dog in France: motorway walks for drivers & dogs (Rees)
)Winston ... the dog who changed my life (Klute)
You and Your Border Terrier – The Essential Guide (Alderton)
You and Your Cockapoo – The Essential Guide (Alderton)

The publisher and author have designed this book to provide up-to-date information and advice regarding the subject matter covered, but cannot accept liability for any damage incurred to people, animals, property, or assets, which may have occurred as a result of following the advice in this book.

First published in English in April 2012 by Veloce Publishing Limited, Veloce House, Parkway Farm Business Park, Middle Farm Way, Poundbury, Dorchester, Dorset, DT1 3AR, England. © Martin Bucksch & Veloce Publishing Ltd 2012. Fax 01305 250479/e-mail info@hubbleandhattie.com/web www.hubbleandhattie.com
ISBN: 978-1-845843-86-1 UPC: 6-36847-04386-5. Original publication © 2007 Franckh-Kosmos Verlags-GmbH & Co KG, Stuttgart. © Martin Bucksch and Veloce Publishing 2012. All rights reserved. With the exception of quoting brief passages for the purpose of review, no part of this publication may be recorded, reproduced or transmitted by any means, including photocopying, without the written permission of Veloce Publishing Ltd. Throughout this book logos, model names and designations, etc, have been used for the purposes of identification, illustration and decoration. Such names are the property of the trademark holder as this is not an official publication.
Readers with ideas for books about animals, or animal-related topics, are invited to write to the editorial director of Veloce Publishing at the above address.
British Library Cataloguing in Publication Data – A catalogue record for this book is available from the British Library. Typesetting, design and page make-up all by Veloce Publishing Ltd on Apple Mac. Printed in India by Imprint Digital Ltd

CONTENTS

Dear dog owner... .... **4**

Prevention. **6**
   Different food/water 6
   Contact with strange animals 6
   Sun and heat 6
   Water 7
   Winter excursions.... 8

First Aid **10**
   Assessing the vital signs.... 10
   Vital bodily functions 10
   Shock – an emergency 13
   Recovery position and
   transportation.... 15

   Resuscitation 17
   Bandaging 19
   How to stop bleeding.... 22

Crisis situations **24**
   Priorities in an emergency.. 24

At-a-glance guide **26**
   Allergic reactions 26
   Insect bites and stings.. 26
   Tick bites 27
   Snake bites... 30
   Foreign bodies 31
   Sunburn... 31
   Sunstroke or heatstroke 32

Hypothermia 34
Frostbite.. 34
Cystitis.... 34
Burns 36
Vomiting 36
Diarrhoea 37
Swallowing foreign objects. 37
Travel sickness 38
Bites and cuts.... 41
Poisoning 44

On the move... 48
What you will need... 48

Foreign travel. 50
Considerations... 50

Chronic illness.... 52
Is the country dog-friendly?. .... ,.... 52
Illness during the journey 53
Becoming familiar with the
unknown.. 53
On the journey 53

Appendices 60
Books 60
Websites 60
English/German/French
Lexicon 61

Index 62

At home and away

Dear dog owner

Whether on a long journey or a day out with your four-legged friend, this guide should help your trip go without any problems as it provides tips on how to prevent emergencies, and information that will allow you to recognise and deal with illnesses and emergencies, should they occur.

Also included is information about what you will need in order to take your dog abroad, matters you should consider before a trip, and everything you will need to take with you, ensuring you are well prepared for any eventuality.

I hope that my book will help ensure that the time you spend with your dog is fun and problem-free for you both!

Martin Bucksch
Veterinarian

Photo credits

The 25 medical photos were taken specifically for this book by Martin Bucksch and Matthias Engelmann. Colour photos from Juniors picture archive (pgs 30, 33, 40, 56); Eva-Maria Kramer (pg 57)/www.infohund.de; Christof Salata.Kosmos (pg 45; Horst Streitferdt (pg 43); Sabine Stuewer (pgs 7, 9); Sabine Stuewer/Kosmos (pgs 38, 55)/www. stuewer-tierfoto.de and Karl-Heinz Widmann/Kosmos (pg 35); Jude Brooks (pgs 42, 45, 58)

EMERGENCY FIRST+AID FOR DOGS

When going on holiday or a weekend trip away, there are a few precautionary measures you can take to make the trip more pleasant for your dog, and ensure that he is not in any danger. Dogs, like humans, can sometimes have difficulty adjusting to a foreign environment: for example, the climate.

Different food/water

If you have to give your dog different food, ensure you introduce it slowly by mixing the new food with the old in gradually increasing amounts. This way you will be able to avoid any stomach and bowel problems which can be caused by a sudden change of food.

Ensure that your dog's food is not left out for too long, especially in warm climates, as it will go off quickly and could lead to an upset stomach.

Avoid letting your dog drink stale water or out of strange water bowls. Also do not feed him anything with bones in as this could be fatal.

Contact with strange animals

Avoid allowing your dog to have contact with native animals, especially strays, and particularly when in countries in the southern hemisphere. Parasites, dermatophytes (fungi) and other diseases can be harmful to your dog. Even if he has been vaccinated, avoid contact with wild animals at all costs.

HIGH RISK OF INFECTION

In countries where dogs are rarely vaccinated regularly – if at all – there is a comparatively high risk of infection as viruses and bacteria that can cause disease and illness are more common and highly contagious. The most important thing to ensure in this case is that your dog has received the correct vaccinations, including those which are not prescribed by law, before arrival in another country.

In some cases, diseases such as canine distemper, hepatitis and canine parvovirus, and those which are zoonotic (contagious to people: leptospirosis, rabies) can be extremely dangerous, and even life-threatening, and also difficult or impossible to treat.

Sun and heat

Even dogs can sustain skin damage from sunburn. Unpigmented (pink-coloured) areas and the nose are especially vulnerable. Use suncream with UVA and UVB protection and a high sun protection factor (SPF). Dogs who do not have much hair on their stomachs, especially those with short, white fur

At home and away

Dogs can also suffer from sunburn, especially if they have pale-coloured skin.

(Bull Terriers, Dalmations, West Highland Terriers) and not much of an undercoat should be kept in the shade whenever possible. A sun umbrella may be useful in this case.

Water

Does your dog love swimming? There's nothing wrong with this as it's an ideal way to cool him and get him exercising. But always ensure the water is not too deep and there are no strong currents, which can be fatal. Even an apparently calm sea can have strong currents and undercurrents, so keep an eye on the ebb and flow of the water.

If you are unfamiliar with the tides, take your dog swimming on a harness with a long lead, or only let him swim in an area that has been designated safe.

SALTWATER

After he's been swimming in the sea, it's a good idea to wash your dog in fresh water as this will help to untangle his fur, and also prevent him from licking all the salt off himself. If he gets salt or sand in his eyes, rinse with clean, fresh water. If he has a skin irritation or inflammation, use an ointment or eye drops containing antibiotics (but not creams containing cortisone on skin calluses).

Don't let your dog drink saltwater as this will make him sick.

SWIMMING POOLS

Don't let your dog near a swimming pool that is not completely full as he will have no way of getting out, should he fall or jump in. The same goes for man-made canals, etc, with steep sides.

EMERGENCY FIRST+AID FOR DOGS

Winter excursions

Depending on the length and thickness of the fur, every dog is more or less well-insulated from heat and also cold weather, and the subcutaneous fatty tissue under the skin also helps with insulation. A dog's capability to regulate his temperature varies with breed, age, and nutrition. As a general rule, dogs without much fur, older dogs, puppies and malnourished animals are at greater risk of suffering from frostbite or hypothermia.

The risk of hypothermia begins at temperatures colder than -10 °C, or prolonged exposure to the cold, especially if a dog's fur is wet. If the dog is completely wet through, he will be at greater risk of hypothermia at temperatures just below freezing.

No swimming!

Do not allow your dog to swim during the winter when it's very cold. If he should jump into water, get him out, dry him off, and get him warm as quickly as possible.

Don't let him walk on frozen lakes or rivers as he may go through the ice, and if an area of ice cannot support the weight of your dog it will certainly not support your weight if you try to rescue him. Only walk on ice when you know that it is thick enough: for example, if it has been opened to the public and approved as safe, but keep in mind that if a large, heavy dog slips on ice, it could have serious, orthopaedic consequences (torn ligament, sprain, strain or bone breakage).

Snow

Never let your dog eat snow because it could cause stomach and bowel upsets, including vomiting and diarrhoea.

Protecting paws in winter

After a walk, remove all clumps of snow and ice from his paws and coat. To prevent clumps of ice and snow sticking to his feet, carefully and gently trim the fur between the pads.

Wash road salt out of his paws with lukewarm water. Dry and cracked paws can be treated with a rich, oil-based moisturiser intended for dogs.

Frolicking in the snow: delicate paws should be well cared for.

EMERGENCY FIRST+AID FOR DOGS

Assessing vital signs

Correct assessment of vital signs and knowledge of resuscitation can save a life, be it human or canine. A death-like pallor is not as obvious with a dog because he is covered in fur, but can be seen on the nose and surrounding area. Extremities such as ears, legs and tail may feel cold, and the eyes may lose their shine, the surface of the eyeball become dry, and the eyeballs appear sunken in their sockets. But be aware that pallid skin and cold extremities may also be signs of shock.

Dull eyes may be apparent for several hours (up to 12) depending on many external and internal factors. Breathing, pulse and function reflexes are all strong signs of life; even if only one of these factors is detectable, life is still present but it may be necessary to attempt to resuscitate the dog.

Vital body functions (see textbox)

BREATHING RATE (NUMBER OF INHALATIONS PER MINUTE)
You can determine this by observing the rise and fall of the chest, or by holding a small mirror to his nostrils (which will cloud when he breathes out).

HEART RATE (NUMBER OF HEARTBEATS PER MINUTE)
You can test the pulse or listen to the heartbeat. Two thirds

Check the pulse on his chest wall, behind the left elbow.

of a dog's heart lie in the left half of the ribcage. Looking for movement on the side of the chest wall will tell you whether the heart is still beating, and the number of beats per minute if it is. This is more informative than feeling for the pulse rate which will not tell you whether the heart is still pumping blood into the

At home and away

outer regions of the body. For confirmation, lay your hand on the dog's chest wall behind his left elbow. Press your fingers on this area to feel the heartbeat.

Pulse rate
Check the pulse rate (not to be confused with the heart rate) by laying a hand on the upper third of the inside of the dog's thigh. It's possible for the heart to beat without pumping blood to the outer regions of the body, in which case, pulse rate will be slower than heart rate, or you may not be able to feel a pulse, in spite of there being a heartbeat.

Circulation
Check circulation by assessing the mucosa (see following) and the so-called capillary refill time (CRT).

The mucosa (inside of the eyelids, gums, muzzle, anus, vulval vestibule) should be pinky-red, smooth and shiny.

To test the CRT, with one finger press on your dog's gums for one second, and then release. There are many small blood vessels called capillaries in the gums, and, when an area of the gum is pressed, blood is forced out of these capillaries. When the pressure is released, the blood should almost immediately refill the capillaries. The time between removing the finger and

✚ Calculating the breathing, heart and pulse rate

You can calculate the rate in each case by counting the number of beats/breaths over 15 seconds, then multiplying the number by four. This will give you the rate per minute.

The pulse can also be checked on the inside of the thigh.

EMERGENCY FIRST+AID FOR DOGS

the return of colour to the gum is called the capillary refill time. You should not press down on the gum for any longer than two seconds as you are starving this area of oxygen, blood supply, etc.

BODY TEMPERATURE
Body temperature can be measured with a normal clinical thermometer. With your dog either standing or lying down, lubricate the tip of the thermometer with a drop of (olive) oil or E45 cream and gently insert it into the anus. Keep the thermometer held downward so that it makes contact with the wall of the anus. If your dog is standing, use your free hand to support his stomach so that he can't sit down.

REFLEXES
Eyelid reflex: As a rule, the eyes should shut automatically if something is about to make contact with the eyeball. Move your finger toward your dog's eye and it should automatically close.

Pupil reflex: If a bright light is shone in the eye, the pupils should contract. To test this, gently hold open an eye and shine a torch into it (not too close), whereupon the pupil should quickly constrict.

Measuring the temperature with a child's digital thermometer.

Toe reflex: Gently pinch the skin between the toes. The leg should draw upward or at least twitch.

If all reflexes are working as they should but you cannot detect

At home and away

✚ Normal rates & measurements for a dog at rest

Heart rate (beats per minute)
60-100 (LARGE DOGS)
90-120 (SMALL DOGS)

Breathing rate (number of breaths per minute)
15-30 (LARGE DOGS)
30-50 (SMALL DOGS)

Mucosa colour & appearance
PINKY-RED, SMOOTH, MOIST

Capillary refill time (in seconds)
1-2

Temperature
38.3-38.7 °CELSIUS (100.7-101.7 °FAHRENHEIT)

breathing or a heartbeat, begin resuscitation without delay (see page 18).

Shock – an emergency

Most serious medical emergencies are followed by a state of shock. Shock in the medical sense is a very serious, life-threatening state because it can lead to complete failure of the circulatory system.

NEUROLOGICAL SHOCK

Serious trauma (eg a large impact) to the central nervous system (the brain and the spine) immediately causes a major reduction in blood flow to the site, which can last for as long as 24 hours and becomes progressively worse if left untreated. Blood vessels in the brain also begin to leak, sometimes as soon as 5 minutes after injury. Cells that line the still-intact blood vessels in the spinal cord begin to swell, and this reduces blood flow to the injured area. The combination of leaking, swelling, and sluggish blood flow prevents delivery of oxygen and nutrients to the neurons (the core components of the nervous system, which includes the brain, spinal cord, and peripheral ganglia), causing many of them to die.

The body continues to regulate blood pressure and

EMERGENCY FIRST+AID FOR DOGS

➕ Causes of shock

Loss of fluid (bleeding, vomiting and diarrhoea)
Allergic reaction/anaphylaxis
Heart disease
Infection
Endocrine disorder and metabolic disease
Poisoning
Over-heating/hyperthermia

heart rate during the first one to one-and-a-half hours after injury, but as the reduction in the rate of blood flow becomes more widespread, self-regulation is not maintained and blood pressure and heart rate drop.

THE TWO PHASES OF SHOCK
Phase 1 – centralisation

If the body experiences substantial fluid loss due to heavy bleeding, vomiting, diarrhoea or burns, blood flow to less important areas is 'switched off' when blood vessels on the surface of the skin, the musculature system, and also the bowel and less important organs narrow. This allows more blood to travel to vital organs such as the heart, lungs and brain.

If the fluid loss is replaced by transfusion during this first phase, the patient will survive and the body will be able to recover.

Phase 2 – decentralisation
Because of the fluid loss, the blood warns the organs by sending out something called a 'messenger substance.' The organs are now suffering a lack of oxygen and the waste products from the cells are no longer being transported away. The blood vessels respond to the signals by expanding, to provide a new supply of blood for the peripheral organs, which means that the vital organs – heart, brain, lungs – do not receive what they need to survive.

Blood pressure then drops suddenly to a life-threatening level, and the heart beats 'empty' without blood in it. Once decentralisation sets in, in most cases it is too late to save the patient.

At home and away

WHAT ARE THE SIGNS OF SHOCK?
- Cold to the touch
- A fast, palpitating heartbeat and quick, shallow breathing
- A longer CRT time; pale, light-coloured mucosa
- Little or no urine, very dark, concentrated urine, blood in urine. A lack of urination over a period of a few hours can be the first sign of shock, also of a renal or bladder problem or even a tear in the urethra
- Apathy/listlessness/lethargy
- Sudden itchiness, restlessness and coming out in hives (if it is an allergic shock)

If you suspect that your dog has gone into shock, you should take him to a vet as quickly as possible, so that he can be given a blood or fluids transfusion to prevent the decentralisation process from beginning. Keep your dog warm (cover him with a rug, blanket or a coat) and try to keep him as calm as possible. Avoid causing him any further stress, such as travelling on busy public transport.

Recovery position and transportation
Lay the injured dog on his side (preferably his right side) so that you can examine him or begin resuscitation. The head

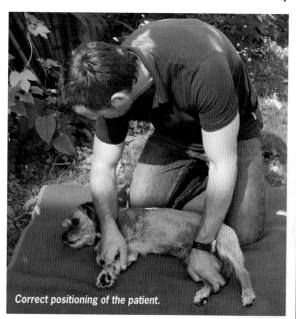
Correct positioning of the patient.

EMERGENCY FIRST+AID FOR DOGS

should be forward and the neck straight so that breathing is not hindered in any way.

TYING A MUZZLE
If your dog is conscious and you need to examine him, put a makeshift muzzle on (a bandage, belt or tie), as pain or stress could cause him to snap. Biting can be a reflexive action, and is a natural defence reaction to pain.

• Ribs
If you suspect a fractured rib(s), lay the dog on the side that has been injured, as this will allow the uninjured side to function better and maintain breathing. Fractures should be splinted, if possible (see page 22).

TRANSPORTATION
The best way to transport your dog is on a stretcher, which you can improvise with a blanket, a board, or something similar. Ideally, you should use a hard base to keep the backbone as straight as possible. If possible, use two people to transport the injured dog. Have the second person drive the car so that you can sit with your dog and ensure he is safely positioned. Avoid extra stress of any kind, so as not to aggravate his condition.

Loop the makeshift muzzle underneath the dog's snout ...

... and tie across the top ...

... take the ends of the muzzle back underneath the dog's nose ...

... then up behind the ears and tie together.

Drive as quickly as you can within legal limits, and do not brake suddenly. Try to keep your composure, in spite of the situation.

Do not play loud music or smoke during the journey (your dog needs oxygen!).

Drive with the windows open to ensure a flow of fresh air through the car.

Resuscitation

Never practice resuscitation (mouth-to-nose, cardiac massage) on conscious dogs which are well, as this could result in life-threatening injuries!

Practical exercises should always be carried out on dummies, so it might be a good idea to enquire about first aid courses where you can practice this.

Emergency First+Aid for Dogs

The ABC of Resuscitation

If your dog has stopped breathing, artificial respiration can maintain the essential supply of oxygen to the blood. Learning the procedure could save your dog's life. If you've ever learnt how to perform the technique on humans, you'll find there isn't a great deal of difference for your dog.

In most cases, artificial respiration and CPR (cardiopulmonary resuscitation) will serve only as a means of keeping a dog alive until a medical professional can treat him. If your dog seems to be having heart or breathing difficulties, contact a vet immediately.

First, identify the problem by following the simple ABC rule: A is checking the Airway, B is for Breathing and C is for Circulation. If the dog is not visually breathing, place your ear on the chest and listen for a heartbeat, or take his pulse.

Open your dog's mouth, grasp the tongue and pull it as far forward as possible, clearing it from the back of the throat.

Wipe away any mucus or blood. Remove any obstruction. Watch your fingers as you could easily get bitten. Make sure

Hold the muzzle firmly shut, extend the neck, and blow air into the nostrils.

At home and away

that you remove the collar and anything else that may restrict air supply.

If the animal has fluid in his throat or is a victim of drowning, hold him upside down by the rear legs for 15-30 seconds, but check for other injuries such as a broken leg before doing so. If you suspect there may be a fracture, hold up the dog by the waist instead: you may need to ask for help if you have a large dog.

Pull your dog's front legs forward so they aren't resting on the chest, making it difficult for him to breathe.

If your dog does not resume breathing once the airway has been cleared, begin artificial respiration. Close his mouth and keep one hand under the jaw for support. Place your mouth over the nose and exhale, forcing air through the nose to the lungs. Be very careful not to exhale too forcefully, as our lungs are bigger and you can run the risk of over-inflating those of the patient.

Watch the dog's chest to see if the lungs inflate, repeating the cycle about six times a minute and being very careful not to inhale saliva or air from the animal.

Continue in this way, giving 20 breaths per minute (one breath every three seconds), until the dog is breathing by himself. To check the heartbeat, move your hand to the lower part of the dog's stomach and then back to around the third or fourth rib.

Bandaging
BANDAGING A PAW

Cover the cleaned wound with sterile muslin or gauze. It's important that you use a covering for the wound that is as

Place strips of cotton wool between the toes.

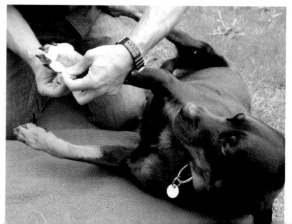

Then wrap the paw in a bandage.

An outer bandage is gently applied ...

sterile as possible, or at the very least it must be clean. Place small strips of cotton wool between each toe; don't forget the thumb and the dew claw.

Starting at the paw, wrap the bandage (preferably adhesive bandage flex which sticks to itself and requires no fixings of any kind) around, ensuring that all of the paw is covered, and extending up over the hock. Don't make the bandage too tight or it will cause more harm than good.

Pressue dressing/tourniquet
The most effective and safest way to control bleeding is to apply direct pressure on the wound with several sterile gauze

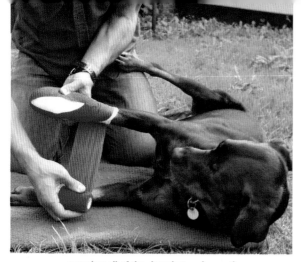

... covering all of the dressing underneath ...

... and then secured with a strip of plaster.

pads, or a folded clean towel. Apply direct pressure for 5 to 10 minutes.

Do not keep lifting the gauze to see if the bleeding has stopped as this disrupts the clot that may have formed. Similarly, if the gauze is soaked with blood, do not remove it since it contains important clotting factors. Instead, keep adding more gauze or towels on top of the soaked ones.

If possible, try to elevate the source of bleeding so that it is above the level of the dog's heart.

If a pressure dressing is not enough to stop the flow or slow it significantly, especially if there is arterial bleeding, a tourniquet may be needed. Never use a tourniquet if the

Emergency First+Aid for Dogs

bleeding can be controlled by direct pressure, and if at all possible, only apply a tourniquet on the advice of a vet, or if you see no other way to control the bleeding before getting the dog to a vet.

Tourniquets are best used on limbs and tail to control arterial bleeding. To apply, find a rubber tube, belt or even a shoelace. Tighten the material above the wound at a point somewhere between the wound and your dog's heart. Loosen the tourniquet every 10 minutes to prevent muscle hypoxia (insufficient oxygen reaching the muscle tissues) and to check for persistent bleeding. If the bleeding has stopped, remove the tourniquet and apply a pressure dressing. If the bleeding continues, let the blood flow for 30 seconds and retighten the tourniquet for another 10 minutes.

Bandaging knee fractures
Visible fractures below the elbow and knee can be protected (you can create a makeshift splint with a piece of wood, or even a magazine) to prevent further slippage of the bone.

The limb should be bound with a bandage, followed by a thick layer of cotton wool which the splint, or makeshift splint, can be laid on top of so that it covers the adjacent joints.

Then wrap the whole thing in another layer of bandage and hold in place with adhesive tape, or use adhesive bandage.

Bandaging the body (chest wall, stomach)
For external injuries to the stomach or chest wall, first ascertain where the wound actually is and cover with clean gauze. Wrap a bandage around the body and apply pressure as necessary to control bleeding. Get your dog to a vet as soon as possible.

Bandaging the head
Injuries to the ears don't heal very easily, and wounds can re-open, so it may be necessary to bandage the head. Hold the injured ear to the head, and if the wound is bleeding heavily, apply an absorbent pad, which should be as sterile as possible, or at the very least clean and lint-free. Finally, apply padding of cotton wool and then bandage around the head to prevent further bleeding should your dog shake his head.

How to stop bleeding
Generally speaking, bleeding is either internal and external (visible), categorised as either light (capillary bleeding) or heavy (venous or arterial bleeding).

Capillary bleeding is the least serious type, and occurs, for example, with a small bite wound, graze or scratch, where

At home and away

no more than a couple of millilitres of blood will be lost. In this case, cold water and a dressing will be enough to bring the bleeding under control.

A steady flow of dark-red blood usually indicates venous bleeding. It's this colour because it does not contain much oxygen as the blood is en route around the body, back to the lungs. For this reason, it's usually fairly easy to control as the pressure is low.

Arterial bleeding, on the other hand, is the least common and most dangerous type of bleeding. Rich in oxygen, it is bright red in colour, and often profuse and spurting due to the high pressure it's under. As so much blood can be lost, it's vital that pressure is applied and maintained until medical assistance is available.

Internal bleeding is possible after a trauma – such as an accident – or if the dog is limp, lethargic or lifeless, has pale mucous membranes (gums, lower eyelids, etc), or is bleeding from any openings (nose, mouth, genitals). Also a sudden increase in abdominal girth (because of a rupture or laceration of an organ within the abdominal cavity) is a serious indication of internal bleeding. If pulse and heartbeat are clearly weak and breathing is shallow, get your dog to the vet as soon as possible.

TREATMENT FOR BLEEDING

Treat capillary bleeding by cleaning the wound thoroughly and then applying a dressing or bandage.

Light venous bleeding can be slowed by splashing cold water on the area and then dressing as for capillary bleeding. If venous bleeding is heavy, apply pressure below the wound and then dress with a pressure pad and bandage, before taking your dog to the vet.

With arterial bleeding, pressure should be applied above the wound and the wound dressed as for venous bleeding. In this case, though, it is essential that pressure is maintained until you are able to get medical assistance for your dog.

Bruises (bleeding under unbroken skin) can be treated with an ice pack, which should cause the blood to disperse eventually.

VETERINARY TREATMENT

Any injury that results in bleeding should be examined by a vet on the day it occurs. A deeper wound that has been thoroughly cleaned will heal quicker with stitches or stapling, and the chances of it becoming infected are considerably reduced if a course of antibiotics is also prescribed.

EMERGENCY FIRST+AID FOR DOGS

Priorities in an emergency
- Keep calm!
- Ensure your own safety so that you are able to care for your dog
- Secure your dog in some way that will not aggravate any injury he may have
- Talk to him gently and calmly to reassure him
- If your dog will allow you to, assess his condition, remove any cause of danger and, if appropriate, begin resuscitation if he is unconscious
- If your dog is reacting aggressively due to fear and/or pain, endeavour to tie a makeshift muzzle around his mouth (see page 16)
- Treat any injuries to the best of your ability
- Organise transportation
- Ring your vet to advise you are on the way
- Carefully transport your dog to the vet

CHECK THE FOLLOWING BODILY FUNCTIONS
- How is his breathing? Is there any indication that his airway is blocked. If yes, try to locate and remove the blockage. Can you hear any unusual noises when he breathes, eg gurgling? Is breathing frequency normal or too quick?

Check the colour of the gums and capillary refill time.

- What colour are the mucosal membranes? Is there a possibility he may have gone into shock? What is the capillary refill time? (See page 11.)
- Check heartrate and feel his pulse
- Is he bleeding from anywhere? If so, how bad is the bleeding

At home and away

and is the blood bright red (arterial) or dark red (venous). Is it pulsating (arterial) or steady (venous)?
- Can your dog stand and/or walk?

IF YOUR DOG IS UNCONSCIOUS
Lay him on his right side (as shown in the picture, right), with his head at a lower angle than the rest of the body, if possible, and check the following:
- Airways for any blockage, and breathing frequency. Listen for any unusual noises when he breathes (gurgling)
- Colour of mucous membranes (gums), and capillary refill time (CRT)

- Heart rate and pulse
- Any bleeding (how badly and which kind)
  If applicable:
- Free airways of blood, mucus, foam or any foreign objects, and, if necessary, perform mouth-to-nose resuscitation
- Attempt to stop any bleeding
- Carry out cardiac massage

Perform cardiac massage directly over the heart region behind the left elbow.

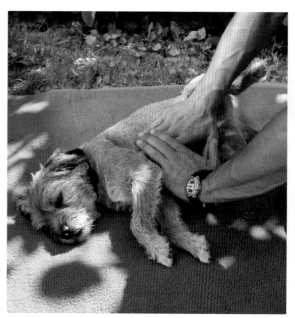

EMERGENCY FIRST+AID FOR DOGS

Allergic reactions

An allergic reaction is the body's way of responding to an 'invader.' When the body senses a foreign substance, called an antigen, the immune system is triggered. The immune system normally protects the body from harmful agents such as bacteria and toxins. Its overreaction to a harmless substance (an allergen) is called a hypersensitivity reaction, or an allergic reaction.

SYMPTOMS
- Itching, swelling, difficulty breathing, circulatory collapse

TREATMENT
If you are aware of any allergies (food, insect bites, medicines, etc) that your dog may have, ensure you always carry medication for him, prescribed by your vet.

For less serious allergic reactions (insect poison or saliva, environmental allergens, such as pollen, dust, mildew etc), cool the area with an ice pack or apply anti-histamine gel or cream (do not allow your dog to lick this). Special dog shampoos containing, for example, oatmeal, can help relieve itching and remove any allergens, such as pollen, which have become caught in the hair.

Insect bites and stings

SYMPTOMS
- Sudden cry of pain, licking and scratching of a particular place on the body
- Inflammation (swelling, redness, soreness) around a sting or bite. If your dog has an allergic reaction to insect poison, this can lead to a severe local or systemic reaction (where the whole body is affected). Hives, also called urticaria, is a skin reaction characterised by pale, slightly elevated swellings (wheals) that are surrounded by a red area and have a clearly defined border. In severe cases, these can cause a dog to go into shock
- Allergic reaction to the chemicals in the saliva of certain insects such as fleas, gnats or bot flies
- Danger of suffocation from a bee, wasp or hornet sting in the throat, mouth or area around the muzzle due to swelling, especially of the mucous membranes

TREATMENT
- If you can see the insect stinger, carefully remove it
- Apply an ice-pack to the swelling, or a vinegar and water solution (half and half)
- Pressing a layer of onion to the affected area will also reduce

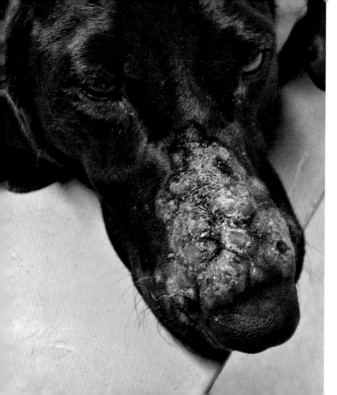

At home and away

An intense and painful-looking reaction to an insect sting.

swelling. but don't use on the eyes, nose or mouth area
- Apply an anti-histamine gel or a cream containing cortisone as prescribed by your vet
- Add a calcium supplement to food to calm the itching
- Put a sock or a T-shirt on your dog so that he can't scratch, lick or bite the area. Even a tiny insect bite or sting can quickly result in a swollen, painful wound

If the sting is in the throat or mouth, or the area around the mouth, and is causing difficulty breathing or hives, get your dog to a vet as soon as possible so that he can be administered a cortisone preparation and/or an anti-histamine to ease the swelling and prevent suffocation. En route to the vet, apply ice to the area to keep it cool and reduce swelling.

Tick bites

Ticks carry diseases that can be dangerous or even life-threatening to both dogs and people. Ticks are parasitic arthropods that feed on the blood of their hosts. They are attracted to warmth and motion, often seeking out mammals

Emergency First+Aid for Dogs

– including dogs. Ticks tend to hide out in tall grass or plants in wooded areas waiting for prospective hosts. Once a host is found, the tick climbs on and attaches its mouthparts into the skin, beginning the blood meal. Once locked in place, the tick will not detach until its meal is complete. It may continue to feed for several hours to days, depending on the type of tick.

On dogs, ticks often attach themselves in crevices and/or areas with little or no hair – typically in and around the ears, the areas where the insides of the legs meet the body, between the toes, and within skin folds.

Symptoms

Though they are known vectors of disease, not all ticks transmit disease – in fact, many ticks do not even carry diseases.

However, the threat of disease is always present where ticks are concerned, and these risks should always be taken seriously. Most tick-borne diseases will take several hours to transmit to a host, so the sooner a tick is located and removed, the lower the risk of disease.

The symptoms of most tick-borne diseases include

Tick bites can have serious consequences if the tick is not removed quickly and completely.

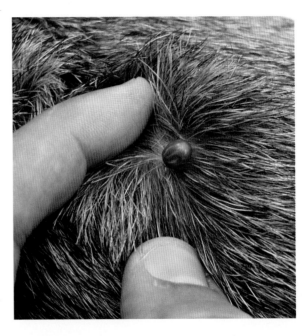

At home and away

fever and lethargy, though some can also cause weakness, lameness, joint swelling and/or anaemia. Signs may take days, weeks or months to appear. Some ticks can cause a temporary condition called 'tick paralysis,' which is manifested by a gradual onset of difficulty walking that may develop into paralysis. These signs typically begin to resolve after the tick is removed. If you notice these or any other signs of illness in your dog, take him to your vet as soon as possible so that proper testing and necessary treatment can begin.

The following are some of the most common tick-borne diseases:

- Lyme disease
- Ehrlichiosis
- Rocky Mountain spotted fever
- Anaplasmosis
- Babesiosis

Prevention
Vaccines against Lyme disease are, as yet, not very effective, but your vet can prescribe a multitude of effective medicines in the form of collars and 'spot' treatments which are safe and effective against ticks. Most products should be applied

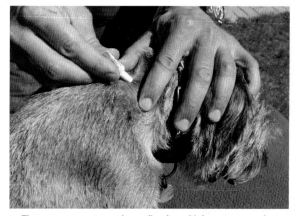

The correct way to apply medication which protects against fleas and ticks.

monthly and used throughout the entire 'tick season,' from around March until November.

Removing the tick
To prevent the possibility of disease, the tick should be

EMERGENCY FIRST+AID FOR DOGS

removed as soon as possible, and it is essential that the mouthpiece is fully removed or else it will remain embedded in the skin and cause infection.

Contrary to popular belief, you should not treat the tick with olive oil, nail varnish remover, alcohol or a lit match! All of these will agitate the tick and cause it to release more infectious fluid from its digestive system into the wound site. And, of course, a naked flame near your dog is not a good idea under any circumstances.

Use tweezers or, even better, a tick hook or tick twister (which you can buy from most pet shops/vets) to clasp the tick tightly, and twist and pull at the same time so that the mouthpiece comes out with the body.

Ensure the tick is properly destroyed so that it cannot attach itself to another animal.

TREATMENT OF THE WOUND SITE

At the wound site, a small, defined red rash or swelling will often appear. If some of the mouthparts remain, although these alone cannot transmit disease, to prevent the chance of secondary infection remove these also. Disinfect the wound site with a solution containing iodine or an antiseptic cream.

Snake bites

It's usually the case that snake bites are rarely seen as they happen, but if your dog is trembling, excited, drooling, vomiting, has dilated pupils, or has collapsed, it's possible that the bite is poisonous. Venomous snakes can leave two puncture wounds at the site of the bite, which is usually on the head or legs. There will also be very noticeable swelling.

TREATMENT
• Keep your dog calm

The bite from an adder can have life-threatening consequences.

At home and away

- Do not cut into the wound and try to suck out the poison as this will only increase blood flow to the region, and speed the spread of the venom
- Apply an ice-pack to the wound site to slow blood flow
- If the bite is on a limb, do not apply a tourniquet
- Apply ice to the wound and tightly bandage it
- **Carry** your dog to a vet as soon as possible (movement will increase blood flow) so that anti-venom medication can be given. Not all vets have this but may know where it can be obtained in an emergency

Foreign bodies

Hordeum murinum, commonly known as wall barley or false barley, is a spiked grass which can find its way into the outer ear of a dog, the foreskin of the penis, under the eyelids, or between the toes, attaching itself by barbed hooks.

Treatment
If you spot this grass on your dog, remove it as soon as possible. You may need to cut it out of the fur, especially if the dog has very long or thick hair.

After walking through high grass, always check between his toes!

Symptoms
An extremely painful and unpleasant reaction to this foreign body are granulomas (a small area of inflammation on the body), or abscesses, especially between the toes, as the heads of the barley penetrate the thin, hairless skin here, working their way right in. Swelling and soreness results, in some cases causing the dog to become lame, leading to a fistula that will become infected with pus.

If your dog gives a sudden cry of pain or holds his head to one side, this can be an indication that a foreign body such as barley grass has got caught in his outer ear, and may already have begun to bury itself in the eardrum.

Treatment
In both cases, unless the grass is very evident and you can safely remove all of it, take your dog to a vet so that he can do so using the correct instruments. It's entirely possible that it will be necessary to anaesthetise your dog to do this, so prevention really is the preferred course of action in this case.

Sunburn

Dogs with pale-coloured skin – for example those with white fur or large areas of unpigmented, hairless areas of the body – are

EMERGENCY FIRST+AID FOR DOGS

at risk of sunburn and the serious illnesses associated with it, such as skin cancer.

SYMPTOMS
Sunburn can be recognised by red, swollen skin, which will eventually peel, blister, and scab.

TREATMENT
- Cool the affected area with an ice pack or similar
- Later apply a gel containing anti-histamine which should help the sunburn to heal. However, if the sunburn is very serious, you should take your dog to the vet

Sunstroke or heatstroke

Dogs cannot lose heat by sweating, since they do not have sweat glands; all they are able to do is pant.

Heatstroke is one of the most common causes of avoidable canine death, so on no account should a dog be left alone in a car, which will quickly heat up and literally cook the dog to death.

SYMPTOMS
- Rapid, heavy panting, salivation
- Staggering or stumbling, collapse
- Fast breathing, quick, weak pulse, body temperature of up to 42°C (very dangerous)
- Red, blue or very pale skin (shock)
- Sickness and/or diarrhoea
- If the brain overheats, this can cause the dog to go into shock

TREATMENT
- Take your dog to a cool, shady place immediately
- If possible, immerse his body in cool – not cold – water, or wrap him in a towel or similar soaked in cold water, pouring more cold water over the towel to keep it cool
- Offer some water (not too cold) for him to drink
- Massage his limbs to get the circulation going again

If your dog lies flat out on the ground and can no longer stand, or is unconscious, he is clearly showing signs of shock, and you must get him to a vet as soon as possible.

Keep the back of his neck and his head cool on the way.

Ensure that you and your four-legged friend both have a happy holiday.

EMERGENCY FIRST+AID FOR DOGS

Hypothermia

Depending on the thickness and texture of the fur, some dogs are better protected from the cold than others.

Hypothermia is most likely to occur if your dog has been in freezing water, even for just a few minutes, but it can also result from shock, after anaesthesia, and in newborn puppies.

TREATMENT
Wrap your dog in a wam blanket/towels/rug and, if possible, take his temperature rectally. If this is below 37 degrees C (98.5 degrees F), get immediate veterinary help (and also if unable to check his temperature).

Keep your dog warm but avoid overheating.

Frostbite

Extreme temperatures can lead to localised frostbite on the extremities such as tips of ears and tip of the tail, paws, testicles and nipples.

SYMPTOMS
The initial signs are areas of numbness, coldness and/or pallor which will later become red, swollen, and sometimes painful.

Serious frostbite is indicated by blisters forming, and areas of the skin dying off (white, numb body tissue).

TREATMENT
Warm the affected parts with tepid water of 32.2 degrees C (90 degrees F). Thawing should occur within around ten minutes, and the skin may appear reddened.

Take your dog to a vet immediately.

Cystitis

SYMPTOMS
Cystitis (inflammation of the bladder) can be caused by a chill – for example, after swimming – and can be very painful.

Symptoms include urinating more often than usual (repeatedly urinating in small amounts, often accompanied by whimpering, which indicates that the dog is in pain). Symptoms such as painful urination or blood in the urine can also be caused by kidney disease. The urine may or may not contain blood.

TREATMENT
Warming the area around the bladder with a hot water bottle

At home and away

His first meeting with the element 'water' ...

... a quick shake and he's on his way!

EMERGENCY FIRST+AID FOR DOGS

fully encased in a cover or towel so that it does not burn the skin can ease the pain.

Give plenty of water to drink, and take your dog to the vet for treatment, which may include pain relief.

Burns

Burns can be caused by getting too close to hot steam, a grill or open fire, or consuming hot food or drink.

SYMPTOMS

A red, swollen area of skin and any obvious signs of pain. The injury could be anything from a first degree burn (a typical burn blister) to charring and extreme damage to the epidermis (third or fourth degree burn), which can also be a cause of shock.

TREATMENT
- First of all, treat any signs of shock (see page 13)
- Cool the burn by holding the area under running water or applying cold water to the area, followed by an ice-pack
- Cover the area with sterile muslin, bandage or even a clean dishcloth
- Do not use ointments or powder!
- Take your dog to a vet as soon as possible

Vomiting

Vomiting is not always a sign of an emergency: sometimes dogs simply eat too much, or eat something that doesn't agree with them, or it may be caused by car sickness or nervousness.

SYMPTOMS

Repeated and/or severe retching or vomiting is almost certainly a sign of a stomach or bowel problem, especially if the vomit contains blood and the dog appears to be in any discomfort (faintness, pain). If this is the case, you should take him to a vet immediately.

A very painful, bloated stomach can be a sign of an acute life-threatening gastric torsion which requires immediate surgical correction.

Common causes of repetitive vomiting, which can be diagnosed by X-ray, ultrasound and blood tests, are –

- Foreign bodies
- Infection
- Parasite infestation
- Liver or kidney failure
- Stomach mucositis or a sudden change of food

At home and away

TREATMENT
- Check temperature and the vomit and faeces for any blood, foreign bodies or parasites
- If there is persistent vomiting and/or blood present in the vomit, get your dog to a vet immediately
- For isolated cases of vomiting, withdraw food for 24 hours but ensure fresh water is available
- If the vomiting subsides over the following few hours, the next day you can feed him with 3-5 small meals which consist of an easily digestible bland diet (rice and chicken or scrambled egg)

Diarrhoea

Since the stomach and bowel are linked in both a functional and anatomical sense, most causes of vomiting can also give rise to diarrhoea as well. The general rule is: every case of diarrhoea must be treated by a vet if it –

- Is explosive and/or painful
- Is black or contains blood
- Continues for longer than two days
- Is affecting the wellbeing of the dog
- Is also accompanied by a high temperature
- Is very watery (which will mean the dog is dehydrated and will need electrolytes)

Simple diarrhoea can usually be treated by fasting your dog for 24 hours, followed by bland food such as chicken/scrambled egg and rice for a couple of days.

Swallowing foreign objects

If your dog has swallowed a foreign object he may pass the object through the stomach and intestines without difficulty, or the object can become stuck in the stomach or intestines causing major problems. Foreign objects may also pose a hazard to the soft tissues of the throat or stomach, or they may become lodged in the throat.

SYMPTOMS
- Foaming at the mouth or increased salivation
- Pawing at the mouth
- Sudden vomiting, possibly with blood
- Sudden diarrhoea, possibly with blood
- Vomiting and diarrhoea
- Sudden fever
- Refusal to eat or drink

EMERGENCY FIRST+AID FOR DOGS

- Inability to keep anything down, including water
- A distended and painful abdomen or belly

If you suspect your dog has swallowed a foreign object, or if you witnessed the event, gently check your dog's mouth and throat to see if the object has lodged itself there, and if it can be removed. If the object cannot be removed make sure your dog is breathing okay, try to keep him as still and calm as possible, and immediately call your vet for further instructions.

Never try to induce vomiting or force-feed your dog water, oil, or anything else in an attempt to flush the object out of your dog without the advice of a vet. If used improperly, all of these actions can cause severe, permanent, and sometimes life-threatening damage to the dog's throat, stomach and intestines.

Do not use medicines intended for human consumption. These might be dangerous for a dog and could cause stomach or bowel problems. The same goes for antibiotics or medicine to stop cramps (eg Immodium) because the chemicals and/or the dosage could be incorrect for a dog.

Travel sickness
As travel between countries has become more frequent and,

Little whirlwind on a trail of discovery: 'puppy proof' your house so that he doesn't accidentally swallow anything he shouldn't.

At home and away

from January 2012, easier to do, along with travel sickness the incidence of diseases introduced from other countries has increased.

If your dog suffers from travel sickness, there are remedies – natural ones and more conventional sorts available from your vet – that can help. Even if the sickness is due to fear, it is still stress-related, which can benefit from B6 vitamins.

There is also a specific herb called Scullcap & Valerian which will help against travel sickness.

To help your dog overcome his anxiety about travelling, sit with him in a stationary car to acclimatise him to the environment. You could even take a favourite toy to play with so that he begins to associate the experience with nice things.

When you've done this often enough for him to be relaxed and happy to jump into the car with the engine off, do the same with the engine running, and then try small trips to favourite places that your dog loves to go to, again, associating the car with something pleasurable.

Disease carriers

Most of the diseases which originate abroad are caused by single-cell blood parasites (eg protozoan which causes Leishmaniasis), but parasites (such as heartworm) can also cause bacterial viruses. They are carried by certain insects such as sandflies or ticks.

It's possible to protect your dog against certain illnesses before you travel; for example, by wearing a special collar some weeks prior to departure that will guard against contracting Leishmaniasis, which is a particualrly nasty disease that can prove fatal. Talk to your vet in good time before travelling so that you can take the necessary precautions for your dog.

If when travelling or later your dog displays any ill-effects for whatever reason, take him to the vet and tell him about where you have been.

Of course, it is essential that your dog is properly vaccinated against rabies. Also ensure that he is microchipped in case he should become lost whilst abroad.

Symptoms

If you notice an obvious decline in his general health (taking into account any tick or insect bites over the last few days/weeks), including fever, pale mucosa, bleeding from the nose, blood in the urine or very dark urine, swollen lymph glands, and any other general signs of illness, take your dog to the vet as soon as possible.

At home and away

Bites and cuts
SYMPTOMS

Four to six hours after an injury, the wound will become contaminated with bacteria (this does not apply to any wounds caused by an operation), from which point in time the bacteria will begin to breed. This then leads to an infection in the wound which can cause skin tissue to die off, or formation of an abscess. Germs that enter the bloodstream could cause septicaemia which is usually accompanied by a high temperature and a considerable decline in your dog's general condition.

TREATMENT

Stop the bleeding (see page 22). If bleeding is light, splash the wound with cold water, cover with clean gauze or similar material and apply pressure.

Visible internal bleeding (eg a bruise surrounded by severe swelling) can be eased by applying cold packs, cold compresses or anything cold you have to hand (bag of frozen peas from the freezer).

For an othaematoma (swelling of the external ear), apply a bandage to the head.

Stop your dog from licking the wound by putting a medical collar (a cone) on him if you have one. This will also prevent the bacteria getting into his mouth and surrounding area.

Once the bleeding has stopped, clean the wound with a sterile saltwater solution or mild antiseptic, but note that disinfectants containing alcohol can burn, which may cause even the friendliest dog to bite! You should not use a hydrogen peroxide solution because the chemicals can deform the edges of the wound and interfere with the healing process.

Cover the wound with a sterile dressing and, if necessary, bandage it.

Take your dog to the vet to have the wound examined. The vet will be able to tell whether or not he will need antibiotics for wound infection or possible blood poisoning.

NEVER UNDERESTIMATE A BITE WOUND!

If you examine a bite wound, in most cases you will only see a tiny perforation in the surface of the skin. However, the bite will also have perforated the subcutaneous tissue below, causing

If you are planning to travel in the southern hemisphere, it is an especially good idea to get some advice from your vet about how to protect your canine friend against disease.

EMERGENCY FIRST+AID FOR DOGS

The severity of a bite injury should not be underestimated: take your dog to a vet as soon as possible.

a little 'pocket' where bacteria can gather. Fur and blood stick to the wound opening, giving the impression that it is healing, when, in reality, the bacteria from the bite is trapped under the surface of the skin where it begins to multiply.

These optimal conditions can lead to extensive local tissue necrosis (where the skin tissue dies off), phlegmon (spreading inflammation), or a large abscess, and possibly even blood poisoning.

This is why even the tiniest bite should be treated with antibiotics or, at the very least, the wound must be cleaned with an antiseptic immediately. Then your vet will be able to decide on any further treatment necessary, depending on its severity.

A dog's nose is very sensitive and should be treated with care.

Nose bleed
A variety of things can cause nose bleed: injury, foreign bodies, tumours, and infections; if the blood is coming out of both nostrils, this could signify liver disease, high blood pressure or poisoning. If the nose bleed lasts for longer than fifteen minutes and the bleeding is heavy, you should take your dog to the vet as soon as possible.

First aid
Keep the nose cool using a cold pack or a frozen bag of peas, wrapped in a towel.

Teeth fractures/teeth falling out
Teeth may be fractured during play or after a trauma. The important thing is to determine whether or not the fracture has made the tooth sensitive or is allowing bacteria into the tooth, which could lead to an infection. In this case, your vet must decide whether to remove the tooth or perform root canal treatment.

It is also possible, in theory, to replace a tooth if it has

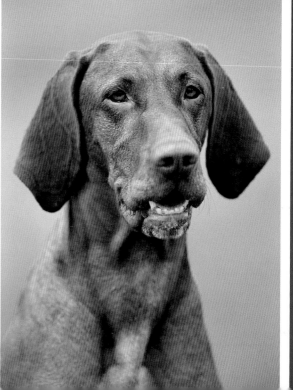

Emergency first+aid for dogs

fallen out, as is the case with people. A tooth will 'survive' for 90 minutes in water or for four hours in UHT milk or an isotonic salt solution.

Poisoning

Certain substances and human foodstuffs can seriously affect the breathing, the skin and, especially in cases of oral intake, the stomach and bowel, and can even be fatal.

Anti-freeze, insect poison, weedkiller, rat poison, dyes, paint and other household chemicals such as tobacco, human medicines and chocolate (especially dark chocolate), and a variety of plants (depending on amount of intake) can cause serious and varied consequences.

Symptoms
- Vomiting
- Drooling, reddening of the mouth, pale mucosa
- Shortness of breath, coughing
- Movement disorder, stumbling, lack of co-ordination

Ensure your dog doesn't get his paws on certain human foods, especially chocolate, which can prove fatal!

At home and away

Watch for signs of listlessness, lethargy, disorientation or sluggishness.

- Apathy, disinterest
- Pain (in stomach)
- Shaking, cramps, shock

Treatment
- Remove the potential source of poisoning
- If he has breathed in the poison, eg carbon monoxide in the

EMERGENCY FIRST-AID FOR DOGS

garage, open the windows or, better still, get him outside in the fresh air
- If the poison is on his skin, bathe the area in clean, lukewarm water to wash it off. If this is not possible then prevent him licking himself by means of a medical collar, a muzzle, or even by covering him in a blanket or T-shirt
- Fat soluble substances, such as tar, can be removed with cooking oil

✚ Vital information for your vet

- Your dog's age, weight, gender, and your telephone number in case the vet needs to call you back
- Everything you know about the poisonous substance
- Try to give an estimation of how much of it your dog has ingested/absorbed
- Try to give an approximate time of when it happened
- What were the initial symptoms that you observed?
- What measures have you taken?

- If he has stopped breathing or his heart has stopped beating, carry out resuscitation (see page 18). Also treat him for shock if necessary (see page 13)

If he has vomited, make sure you keep his airways clear, especially if you notice that he is disorientated or losing consciousness.

In the case of oral poisoning, first check that he has not swallowed any sharp objects, then give him a salt solution so that he vomits up the poison (3 teaspoons of salt in a glass of water). Caution: Only use these measures if the dog is co-operative and fully conscious and you are unable to get him to a vet. If you are in any doubt, then do not attempt this.

MEASURES TO TAKE

If you suspect your dog may be suffering from poisoning, phone the vet for advice straight away. Most cases of poisoning require urgent treatment. If oral poisoning is treated in time, it is possible to remove most of the poison by using a substance to induce vomiting, or a gastric lavage (stomach pump).

If possible, take a sample of the poisonous substance with you (even if in the form of vomit) so the vet knows exactly what he is dealing with.

Puppies will be into everything!

EMERGENCY FIRST+AID FOR DOGS

What you will need
- Antibiotic ointment and eye ointment. Do not use any creams containing cortisone as this can interfere with the healing process and weaken the body's natural defence system
- A mild antiseptic solution for cleaning wounds
- Anti-histamine (tablets or injection) for any known hypersensitivity to allergens (see allergic shock on page 13)
- Burn ointment
- Eye pipette
- Electrolyte powder to replace lost fluids after vomiting or diarrhoea (available from your vet)
- Isotonic salt solution (sterile) suitable for cleaning wounds or flushing out any foreign bodies in the eye
- Thermometer
- Adhesive plaster (a roll)
- Sterile gauze bandages and padded bandages, available from a chemist or your vet
- Cotton wool, ideal for padding out bandages
- Paraffin oil
- Tweezers
- Scissors (sharp) and bandage scissors with a rounded tip to minimise risk of injury
- A small spray bottle to clean out the eyes or any wounds
- Torch
- Tick hook
- Disposable gloves and possibly even a foil rescue blanket
- Bach Rescue Remedy or similar
- Any medication as prescribed by your vet

AND ALSO POSSIBLY ...
- One or two dog booties to protect a paw injury
- A muzzle, for your own safety

Right: Ready for anything: have first-aid kit, will travel ...

Far right: ... and a few treats won't go amiss!

At home and away

EMERGENCY FIRST+AID FOR DOGS

Considerations

Before considering whether or not to take your dog abroad, it is strongly recommended that you talk with your vet to determine if this is the best course of action for your dog, and what will be required for you to do so safely and legally.

As of 1 January 2012 changes to the Pet Travel rules affect entry or re-entry to the UK, as announced by The Department for Environment and Rural Affairs (DEFRA). The statement on the DEFRA website says:

"Pet Travel rules change on 1 January 2012 when the UK brings its procedures into line with the European Union. From this date all pets can enter or re-enter the UK from any country in the world without quarantine provided they meet the rules of the scheme, which will be different depending on the country or territory the pet is coming from."

These changes will make travelling with pets cheaper and easier. The rules on rabies vaccination for entry into the UK will be in line with the rest of Europe. Previously, the UK, and four other Member States were able to enforce additional controls.

If you wish to return to the UK with your pet after a trip abroad, or intend to bring your pet into the UK for the first time, you need to be aware of the changes to the UK pet entry rules that take effect on 1 January 2012.

EU Pet Passport.

PET PASSPORTS UK PET TRAVEL SCHEME CHANGES JANUARY 2012

So what do you and your vet have to do to bring your pet dog, cat or ferret into the UK? Rules depend on which country

At home and away

you are travelling into the UK from. The countries are broadly defined as follows:
- EU member states and approved non-EU countries
- Non-approved countries (unlisted non-EU countries)

For a detailed list of countries in each category, please visit 'Countries and territories' on the DEFRA website (http://www.defra.gov.uk/wildlife-pets/pets/travel/).

Requirements for EU member states and approved non-EU countries
- Have your pet microchipped
- Have your pet vaccinated against rabies. A period of 21 days must then elapse after the rabies vaccination date before entry to the UK is permitted
- Obtain a valid Pet Passport or equivalent documentation (depends on country)
- Dogs must have tapeworm treatment. The treatment must be administered 1-5 days before arrival in the UK
- Arrange for your animal to travel with an approved transport company on an authorised route. For details regarding approved transport and routes please visit 'Routes and transport companies' on the DEFRA website (http://www.defra.gov.uk/wildlife-pets/pets/travel/)

The 1 January 2012 change to the Pet Travel rules means that if you are travelling into the UK from EU member states and approved non-EU countries, the above replaces the need for a rabies blood test and the previous 6 month quarantine period.

There is no mandatory requirement for tick treatment.

Requirements for non-approved countries (unlisted non-EU countries)
- Have your pet microchipped
- Have your pet vaccinated against rabies
- Arrange a blood test at least 30 days after vaccination to make sure the vaccine has given protection against rabies. The length of the waiting period before entry to the UK is three months from the date of a blood sample that proves protection
- Obtain official pet travel documentation
- Dogs must have tapeworm treatment. The treatment must be administered 1-5 days before arrival in the UK
- Arrange for your animal to travel with an approved transport company on an authorised route. For details regarding approved transport and routes please visit 'Routes and

EMERGENCY FIRST+AID FOR DOGS

transport companies' on the DEFRA website (http://www.defra.gov.uk/wildlife-pets/pets/travel/)

The 1 January 2012 change to the Pet Travel rules means that if you are travelling into the UK from unlisted non-EU countries, the above replaces the 6 month quarantine period.

SUMMARY OF CHANGES
For entry into the UK, all pets will still need to be vaccinated against rabies. However, pets from the EU and listed non-EU countries will no longer need a blood test and will only have to wait 21 days before they travel. Pets from unlisted non-EU countries will be able to enter the UK if they meet the criteria to ensure they are protected against rabies, including a blood test 30 days after rabies vaccination, followed by a 3 month wait before entering the UK.

The foregoing is a summary only of the basic requirements for the most common scenarios and the Pet Travel changes introduced. For more details visit and carefully read the information given on the DEFRA website before travelling with your pet (http://www.defra.gov.uk/wildlife-pets/pets/travel/).

(Pet travel information reproduced courtesy of D for Dog: http://www.dfordog.co.uk)

RECOMMENDED VACCINATIONS
Further injections are strongly recommended, or may even be a requirement – depending on the country and/or time of year you travel – such as those for canine distemper, leptospirosis, hepatitis C, canine parvovirus, borreliosis, kennel cough, etc.

HEALTH CERTIFICATE
In some cases, your dog may need a health certificate from your vet and a certificate which states he has been treated for tapeworm and ticks (Great Britain (tapeworm only), Norway and Sweden). Check with your vet or the DEFRA website.

Chronic illness
If your dog suffers from a chronic illness, such as a heart problem, an allergy or a metabolic illness, discuss any potential problems with your vet before taking your dog abroad.

Also ensure that the sort of holiday you are taking is really suitable for your dog; for example, a trekking holiday in the mountains is not ideal for a 15-year-old dog with a heart problem!

Is the country dog-friendly?
Before you plan your holiday, determine whether or not the

At home and away

country is dog-friendly. In some, dogs are not allowed in hotels, and trying to eat out with a dog can be problematic. Your local travel agency can help you with information on this, as well as numerous websites that deal specifically with this subject. Some countries – France and Britain, for example – do not allow entry of certain breeds of dogs, or place restrictions on certain breeds.

Illness during the journey

For dogs that are especially nervous or anxious, or have a tendency to suffer from motion sickness, ask your vet about medication to help with this. You could also try the perhaps more natural Bach Rescue Remedy to help with nervousness and anxiety. This has fewer side effects than sedatives and anxiety medication, and is also useful in other stressful situations or after an accident. It's a good idea to have a trial run with any sedatives prescribed by your vet before you go away.

It is also advisable to chat with your vet about any worries you may have regarding the length of the journey, destination, time of year or type of travel.

Always have your vet's telephone number to hand, and don't forget your pre-prepared, portable dog pharmacy!

Becoming familiar with the unknown

New situations or objects are stressful for dogs, so it's a good idea to acclimatise him beforehand to anything new you may use on your journey (muzzle, lead, dog carrier), or to travelling in the car. With patience, praise and rewards, your dog will soon come round to the idea!

And actually, it's best to familiarise him with these situations right from puppyhood. If he regards his transport cage as a cosy bolthole, he will not mind being in it too much during the journey.

On the journey
TRAVELLING BY CAR

In Britain, as yet, there are no laws to say how and where a dog or dogs should be transported in a car, and neither is there a legal requirement for your dog to wear a harness that can be attached to a seat belt, although this is an extremely sensible idea for the safety of your dog and you/your passengers. The law does require that nothing in the car should distract the driver and that loads should be safely carried. But, leaving aside the legal issues, it's a sensible idea to protect your dog, wherever you go, and a harness, dog guard or even a metal travelling cage will help to keep him safe.

EMERGENCY FIRST+AID FOR DOGS

FEEDING
You should feed your dog at least 12 hours before the start of the journey and, if it's a long journey, feed him little and often en route.

Make sure he has plenty of water available to him at all times.

BREAKS
Take regular breaks so that your dog can get out of the car, move around/relieve himself, and have a drink of water.

Avoid travelling during the hottest hours of the day in the summer: travel early in the morning or in the evening.

Old or dogs in poor health can be especially sensitive to heat.

SUN
Do not leave your dog alone in the car during warm weather. Even a short time can be complete misery – and potentially fatal – for a dog if he overheats. Even if the window is open slightly, not enough air will circulate.

Remember, also, that even though your car may be in the shade when you leave it, the position of the sun will change during the day!

AIRSTREAM
Does your dog love to stick his nose out of the window while you are driving? Be wary about allowing him to do this as it can lead to conjunctivitis. Much better to stop and enjoy the fresh air in a picnic area!

MOUNTAINS
if you are holidaying in or visiting a mountainous area or driving up steep hills, giving your dog a treat to chew on could help prevent earache by releasing pressure build-up in his ears.

DOG CARRIER
For all dogs transported in the hold of an aeroplane, carriers especially for this purpose are available to purchase. When you buy a carrier, make sure your dog can sit, stand and lie in it comfortably, and can also turn around. Make sure your dog is used to his carrier before you travel, perhaps by feeding him in it so that he associates it with something nice. Label the carrier with the words 'living animal,' indicate which way up it should go, and you could also add a small personal message such as "My name is Pepe, I am a little bit afraid of being in this box. Please handle me with care. Thank you." This should help to make the journey as pleasant as possible for your

At home and away

The size of the dog carrier should be appropriate to the size of the dog.

dog. Securely fix a clear plastic folder on the top of the box which contains your name, address, a contact number (your's and your vet's), and the address of your destination. Include photocopies of health certificates and a copy of your dog's Pet Passport. Provide a spare lead in case the flight staff need to take your dog out of his carrier. Place a thick towel in the carrier in case he is sick or needs to urinate. Ensure your dog has access to water at all times; use a special, non-spill bowl for this purpose.

Travelling by train

It's generally possible to travel with your dog on a train. Small dogs (up to the size of a domestic cat) can be carried in a dog carrier for no charge, but larger dogs will incur a charge (as with international travel, this will be the same price as a second class ticket for a child). Large dogs must be on a lead and wear a muzzle (with the exception of guide dogs or assistant dogs for the disabled).

You should bring plenty of water for the train journey as well

EMERGENCY FIRST+AID FOR DOGS

Travelling by train – no problem if you are well prepared!

as a water bowl. If it is a very long journey, consider making a couple of stops to allow toilet breaks for your dog.

Travelling by plane
In order to keep the journey as short as possible, book non-stop, direct flights. Early morning or late evening flights are preferable as the dog carrier may be placed in the entrance of the hold in direct sunlight which could lead to overheating. If possible, book the flight on a weekday when it is usually less busy.

Reservation
Your local travel agent will be able to advise about prices, regulations, etc, or you could ask the flight company before you book. Since the number of animals permitted on each plane is limited, book ahead to register your dog. Some airlines allow small dogs (5-8kg, including the dog carrier) into the cabin. The same goes for guide dogs and dogs for the disabled. If carried in the hold, excess baggage charges apply.

Collection
Find out from the flight company beforehand where you can collect your dog after the flight. Don't forget to have your papers ready and also a photo of you with your dog so that you can prove he is yours, in case of any mix-up!

Travelling by sea
If you are going to be travelling on a ferry or ship, make sure you know exactly what the regulations are with regard to your dog. Depending on the length of the journey, your dog may have to be kept in a carrier or remain in the car, and you may have only limited access to him during the trip. Some companies provide a deck for 'walkies' whilst others do not allow any animals aboard.

Walking/trekking/cycling
The same rule applies to dogs as well as people here: avoid over-exertion, especially if you or your dog have circulation problems (heart disease or metabolic illness). Watch out for

Travelling by ferry: make sure you find out what the company's regulations are with regard to carrying dogs.

EMERGENCY FIRST+AID FOR DOGS

signs of tiredness, shaking, stumbling, heavy panting, cramps, breathlessness, or even fainting. All of these symptoms could signify something more serious! Take plenty of water and something for your dog to eat.

Risk of falling
Keep your dog on a lead if there is a risk of falling, especially if he is being a bit skittish and not paying attention. In most national or nature parks, dogs must be kept on a lead to protect the flora and fauna.

Useful items for your journey
- Your vet's telephone number: Always keep this handy, saved in your mobile phone or on a little card in your wallet. Also note the emergency number and the number of a vet in the area you will be travelling to
- Vaccination papers and necessary medication: if your dog has any allergies, or is diabetic or epileptic, etc, ensure you have sufficient stock of the necessary medication
- Pet insurance certificate: take a copy of this with you
- Muzzle and lead: in some countries or on some means of transport, both are required. They could also be vital in an emergency situation

Lots of people love to go on holiday with their dogs – and their dogs enjoy it, too!

- Food: if you feel there may be a problem with finding suitable food for your dog, take enough for whilst you are away, so that he does not have to adjust to a different type of food, which could cause tummy upsets

At home and away

➕ A little tip ...

A tip I've learned from experience is to take some dog shampoo when I travel with my canine pal. My dog loves to roll in all sorts of unmentionable things so, if possible, I try and get him cleaned up before returning to my holiday accommodation

- Portable water carrier and bowl: essential to have at all times
- Antiparasitics: ticks, fleas and biting insects such as sandflies are carriers of many different dieases that can be difficult to treat or even fatal (see page 26). The further you venture into the southern hemisphere, the more dangerous these diseases can be. Treat your dog with effective, preventative antiparasitic medication before you go and take a supply of it with you

Keep your canine first aid items in a carry case which is easily accessible and labelled appropriately. Check the medicines every 3-6 months to confirm they are still in date and that the contents are still usable.

Ready and prepared!

Emergency first+aid for dogs

Books

Walking the Dog – Motorway walks for drivers and dogs
by Lezli Rees. Hubble & Hattie. April 2011. ISBN 9781845841027

Walking the dog in France: Motorway walks for drivers and dogs
by Lezli Rees. Hubble & Hattie. July 2012. ISBN 9781845844295

Dogs on Wheels: Travelling with Your Canine Companion
by Norm Mort. Hubble & Hattie. August 2012. ISBN 9781845843793

Dogfriendly.Com's United States and Canada Dog Travel Guide: Dog-Friendly Accommodations, Beaches, Public Transportation, National Parks, Attractions
by Tara and Len Kain. Dogfriendly.com. May 2010. ISBN 9780979555107

The Traveling Dog – Tips & Advice If You Plan To Travel With Dogs (Kindle Edition)
by Brian Hill. Javelina House Publishing. November 2008. ASIN B001KBZZLW

The Dog Travel Guide – How To Travel With Your Dog (Kindle Edition)
by Mike Riley. Espino Enterprises Inc. February 2010. ASIN B0037KMGRO

Travel Tips for Dogs and Cats
by David Prydie. Interpet Publishibg. December 1999. ISBN 9781860541186

Dog-friendly Breaks in Britain (Special Places to Stay)
by Alastair Sawday. Alastair Sawday. November 2011. ISBN 9781906136604

Pet Friendly Places to Stay 2012
AA Publishing. October 2011. ISBN 9780749571429

Websites

http://www.defra.gov.uk/wildlife-pets/pets/travel/
www.britishairways.com/travel/pet/public/en_gb
www.passportforpets.co.uk/dog-travel.php
www.doggiesolutions.co.uk/dog-car-travel-5813-0.html
dogs.about.com/od/travel/a/travelwithdogs.htm
www.shipyourpet.com/
gouk.about.com/od/tripplanning/qt/DogtoUK.htm
www.thekennelclub.org.uk/item/1074
www.dog-car-sickness.co.uk/
http://www.racshop.co.uk/travel-touring/pet-travel-accessories.html
http://canineconcepts.co.uk/en/blog/59-why-is-my-dog-getting-travel-sickness
http://www.pet-insurance-comparison.net/dog-car-safety-travelling-with-dog.html
http://www.jamescargo.com/livestock_transport/dog_travel_tips.htm
http://europa.eu/travel/pets/index_en.htm

At home and away

English/German/French lexicon

English	German	French
Accident	unfall	accident
Allergy	allergie	allergie
Bandage (to change)	verband (wechseln)	changement bandage
Bee sting	bienenstich	piqûre d'abeille
Bleeding	blutung	hémorragie
Blood loss	blutverlust	perte de sang
Blood test	blutuntersuchung	analyse de sang
Car accident	autounfall	accident de voiture
Chronic heart disease	herzerkrankung	chronique cardiopathie
Collapse	zusammenbrechen	écroulement
Cut/incision	schnitt/schnittverletzungen	entaille/incision
Diabetes	diabetis	diabéte
Diarrhoea	durchfall	diarrhée
Difficulty breathing	atembeschwerden	respiration difficile
Dog bite	bissig	chien morsure
Electric shock (to get)	stromschlag	recevoir une décharge
Emergency	notfall	urgence
Emergency clinic	notfallklinik	clinique d'urgence
Emergency service	notdienst	service d'urgence
Fall (to have a)	sturz	faire une chute
Foreign body	fremdkörper	corps étranger
Heatstroke	hitzeschlag	coup de chaleur
Hit by (car, motorcycle)	angefahren von auto/motorrad	se faire renverser par une voiture/motocyclette
Hot (overheat)	überhitzt	faire trop chauffer
Injection	injektion	injection
Injury	verletzung	blessure
Insect bite	insektenstich	piqûre d'insecte
Intolerance	unverträglichkeit	intolerance
Pain	schmerz	souffrir
Permanent medication	dauermedikation	prendre des médicaments
Pill	pille	pilule
Poison	gift	poison
Seizure	anfall	attaque
Snake/snake bite	schlange/schlangenbiss	serpent/mordure de serpent
Sunstroke	sonnenstich	insolation
Swallow	verschlucken	avaler
Swelling	schwellung	enflure
Tablet	tablette	comprime
Tick	zecke	tick
Trauma	trauma	traumatisme
Unconscious	bewusstlos	sans connaissance
Veterinarian	tierarzt	veterinaire
Vomit	erbrechen	vomir
Wasp sting	wespenstich	insecte guepe
Wound	wunde	blessure
X-ray	röntgen	radiographie

EMERGENCY FIRST+AID FOR DOGS

INDEX

Aeroplane travel 56
Airway 18, 24, 25, 46
Allergic reaction 26, 52
Antibiotics 7, 23, 38, 41, 42, 48
Anxiety 39, 53
Arterial bleeding 22, 23
Artificial respiration 18

Bandaging 19
Bites & cuts 41
Bleeding 20-24, 39, 41, 43
Blood pressure 13, 14, 43
Blood test 51, 52
Bodily functions 24
Body, bandaging 22
Body temperature
Breathing 15, 18, 19, 23-27, 32, 38, 58
Breathing rate 10, 11, 16
Bruises 23, 41
Burns 36

Capillary bleeding 22, 23

Capillary refill time (CRT) 11, 12, 15, 24, 25
Car travel 53
Cardiac massage 25
Chocolate 44
Chronic illness 52
Circulation 11, 13, 18
Climate 6
Cycling 57
Cystitis 34

DEFRA 50
Diarrhoea 8, 14, 32, 37
Disease 6, 28, 39, 41, 59
Dog carrier 53-56
Dog-friendly 52, 53
Dog shampoo 26, 59

Electrolytes 37
Emergency 13, 24, 36, 58
Exercise 7
Eye ointment 7, 48

Falling 58

First aid 10
Food 6, 54, 58
Foreign bodies/objects 25, 31, 36-38, 43, 46, 48
Foreign environment 6, 50
Frostbite 8, 34

Gastric torsion 36

Harness 53
Head, bandaging 22, 41
Health certificate 52
Heart rate 10, 11, 24, 25
Heartbeat 23
Heatstroke 32
Holiday 6, 58
Household chemicals 44
Hydrogen peroxide 41
Hypothermia 8, 34

Illness 6
Immune system 26
Infection 6, 36, 41, 43
Inflammation 26

At home and away

Insect bites & stings 26, 27
Internal bleeding 23, 41

Knee fracture, bandaging 22

Leishmaniasis 39
Lymph glands 39

Microchip 39, 51
Mucous membranes 23, 25, 26, 39, 44
Muzzle 16, 18, 24, 46, 48, 53, 55, 58

Nose bleed 43

Othaematoma 41

Parasites 6, 36, 37, 39
Paw, bandaging 19
Paws 8, 48
Pet Passport 50, 51
Pet Travel Scheme 50, 52
Poisoning 41-46
Pressure dressing 20, 21
Pulse 10, 11, 23-25

Rabies 39, 50-52
Recovery position 15
Reflexes 12
Resuscitation 10, 15, 17, 18, 46

Saltwater 7, 41
Sandflies 39, 59
Sea travel 57
Sedatives 53
Septicaemia 41, 42
Shock 10, 13-15, 32, 34, 36, 46
Skin cancer 32
Snake bites 30
Snow 8
Southern hemisphere 6, 41, 59
Strays 6
Stress factors when travelling
Suffocation 26, 27
Sun 6, 54
Sunburn 6, 7, 31, 32
Sunstroke 32
Swimming 7
Swimming pools 7

Tapeworm 51
Teeth 43
Temperature 8, 12, 34, 37, 41
Ticks 27, 28, 39, 51, 59
Tourniquet 21, 22
Train travel 55, 56
Transportation 15
Travel sickness 38, 39, 53
Trekking 52, 57
Tumour 43

Vaccinations 6, 39, 50, 52, 58
Venous bleeding 22, 23
Vital signs 10
Vomiting/vomit 8, 14, 32, 36-38, 44, 46

Walking 57
Water 6, 7, 35, 36, 38, 41, 54, 55, 58, 59
Winter 8

Out and about with your dog? Then these are the books you need!

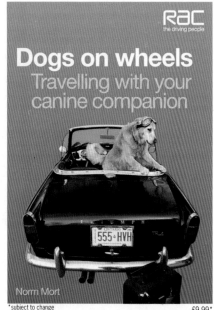

Norm Mort

*subject to change

£9.99*

Looks at you, your family, your vehicle, and most importantly your dog, and tells you how to get the most out of travelling with your four-legged friend – whether for five minutes or five hours.

Helpful advice, insights into your dog's world, and guidance on choosing the right vehicle are only a part of this comprehensive look at travelling with your dog.

Packed with original colour photographs, and containing invaluable information and opinion from veterinarians and an animal behaviourist.

Visit our website (www.hubbleandhattie.com) or call (01305 260068) for full details. P&P extra

"Approved by the RAC ... helpful details such as nearby facilities, contact info for dog-friendly pubs and eateries, and activities for children ...never leave home without this guide!" – Your Dog magazine

"A brand new app is now available with additional features such as maps which track movement and GPS coordinates for sat nav use" – Your Dog magazine

Just £4.99* each, these invaluable guides have the low-down on all the best places to take a break, give the kids a chance to let off steam, and walk your dog when travelling on motorways in the UK and France.
Don't set off without your copy!